THE SQUARE OF SEVENS

AN AUTHORITATIVE SYSTEM OF CARTOMANCY

WITH A PREFATORY NOTICE

BY E. IRENAEUS STEVENSON

TO JOHN DAVIS ADAMS

this new forth-setting of an old mystery is cordially offered.

Editorial Preface

"'Tis easy as lying."—*Hamlet*

It is safe to presume that even the most inquisitive book-hunters of the present day, and few of the fellowship during two or three generations past, have encountered the scarce and curious little volume here presented, as in a friendly literary resurrection—Robert Antrobus's "The Square of Sevens, and the Parallelogram." Its mathematical title hardly hints at the amusement that the book affords. With its solemn faith in the gravity of its mysteries, with its uncertain spellings and capital-icings such as belong to even the Eighteenth Century's early part, with its quaint phrases and sly observations (all the time sticking strictly close to business), it has a literary character, as well as me occult, that is quite its own.

Fortune-telling with cards and belief in fortune-telling with cards—like a hundred greater and lesser follies of the mind—were straws floating along the current of British life, intellectual and social, during the reign of George the Second. This was the case, in spite of the enlightening influences of religion, science, and philosophy. Modish society was addicted to matters over which argument was hardly worth while—in which respect we find modish society the same in all epochs. Our ancestresses particularly were often charming women, and almost as often sensible women; but, like the men of Athens, they were too superstitious. Often were they such in a fond and amusing degree. Lady Betty or Lady Selina—for that matter, even Sir Tompkin and my lord Puce—might be spirited men and women of the world. But they did not repudiate the idea of ghosts. They abhorred a mirror's breakage. They disliked a Friday's errand. They shuddered over a seven-times sneeze or at a howling dog at midnight.

And the gentle sex, especially, would and did tell fortunes almost as jealously as play quadrille and piquet. Let us be courteous to them. Let us remember that Esoteric Buddhism, Faith Healing, and Psychic Phenomena were not yet enjoying systematic cultivation and solemn propagandism; and that relatively few dying folk were allowed to "go on with their dying" as part of a process of healing which excludes medicine and insists on the conviction that the invalids are not ill!

But to our "Square of Sevens"—with which even a Gallio may deign to be diverted—especially if in using it the air is found to be full of coincidences. The story of the book is already alluded to, as odd. The inquisitive reader may be referred to "certain copies only." Therein, "inserted by Afterthought on the Author's part" (and therefore in a mere fraction of whatever represented the extremely small edition of the work), may be sought the "Prefatory Explication, made for the Benefit of My Friends, Male and Female." In recounting the origin of the manual, its author is candid, but at the same time too long-winded for quoting entire. Enough to say, as the substitute for a lengthy tale of facts, that prior to the year 1731 the author of "The Square of Sevens," Mr. Robert Antrobus, "a Gentleman of Bath," was called in the month of November to pass sundry months in Tretelly, that antique but still lively little town of Cornwall. He describes himself as "exceedingly vexed and inconvenienced by Summons on my Affairs connected with the Parcelling of a piece of Property, unexpectedly acquired." Mr. Antrobus—who, by-the-bye, may perhaps be associated in the memories of readers of minor Eighteenth-Century correspondence with such notables of the day as William Pitt, Dr. Johnson, Admiral Byng, Mark Akenside, William Pulteney, the Duke of Cumberland, and many others of the time—was a shy, silent man of wealth. Also was he one of considerable learning, out of the way and other, including an interest in gypsies and gypsy language remarkable for the period.

He lodged at "the only Inn of any suitability" in the place. Thereby be made an unexpected acquaintance. Before a week had elapsed, he became much interested in the fact that under the same roof, but in more bumble quarters than his own, was lying and dying another stranger in the place. This was a man of some forty years, known only as "Mr. George." His home is not a clear matter, nor that he had any relatives except a little girl of six or seven years old, his child. It is likely that in alluding to him in the "Prefatory Explication" mentioned, Mr. Antrobus disguised what was already obscure, and that "Mr. George" of the "troublesome Talk of the Inn-people" is an abbreviated pseudonyme.

Mr. Antrobus was a humane and benevolent man, as well as an inquisitive one. He delicately assisted to make the sick guest more comfortable in his wasting body. He won his confidence, genuinely compassionated his anxieties, and presently pledged himself to a most kindly office—the care and provision in future for the child soon to be fatherless; long before this time motherless. Whether she was motherless by the actual death of the parent, or not, Mr. Antrobus did not learn, or does not tell. But he did learn, by a confession, that "Mr. George" was really George X—, a gypsy, and one withal of unusual education and breeding. More remarkable still, he was a gypsy intensely embittered against' a race from which he had lived for many years wholly withdrawn. The cause of such sentiments and renegade existence good Mr. Antrobus "tryed in vain, with much Delicacy" to discover. At the clearest, it appeared to him to date from the dying man's marriage and from some stormy period of his career. In any case, the renunciation of "Mr. George" in lot and part in gypsydom was of savage sincerity. He would not tolerate the idea of his child being left open to such influences; and, as a matter of her happy fortune in meeting with our kind Bath antiquarian, she never encountered them.

Recognising in his benefactor not only a generous man, but one genuinely interested in the whole topic of gypsy life, character, and affairs (moderately studied at the time preceding a Borrow or a Leland), "George X—" entertained Mr. Antrobus "for hours and dayes" in what must have been an extraordinarily free parliament. It discussed not merely the concerns in general, but the secrets, of Egypt. "Mr. George" bad travelled much. He bad acquired a deal of special knowledge delightful to Antrobus. It is provoking that Antrobus did not commit more of it to paper. But, among other matters, Mr. Antrobus was enlightened on the secrets of looking into dukkeripens *in a degree of minuteness that few gorgios enjoy.*

As part of this last confidence—the rarest from one of the Blood—did George X— disclose in course of certain séances the "Square of Sevens," that most particular and potent method of prying into the past and present and future. In it figures the wonderful "Parallelogram," with its "Master Cards," "Influences," and so on—which our book records. Moreover, George X— declared that whereas most of his race can or will use only corrupted or quite frivolous versions of it, this statement set its real and rare self forth with the utmost purity, value, and completeness, in a degree "known to only a few of all the families of Egypt." As such a weighty bit of Black Art did Mr. Antrobus make its details into a book. As such he printed it. Doubtless he thought that a betrayed secret may lawfully be re-betrayed as fully as possible.

Nevertheless, it was not so much of a re-betrayal. For less than what a publisher of this day would call one fair-sized edition of "The Square of Sevens," printed for Antrobus by the great John Gowne, of The Mask book-shop, has ever appeared. And, to account for the semi-privacy surrounding the little work, must be set forth the dolesome incident of a printing-house fire burning, "all except about a dozen or so of copies," before there had been any "distribution of the Book" among the author's "Friends, Male or, Female, or to the Publick." By some sudden change of his own mind or his conscience, Mr. Antrobus did not order any new edition. The prefatory "Afterthought" mentioned may be found, only if stuck in some of the copies of the volume—doubtless by quick and clumsy after-pastings.

Why Antrobus did not give the volume real currency is not known. That he was urged to do so is certain. It is likely, however, that about this same time some pecuniary losses withheld him from such expensive bobbies as printing books. He returned to Bath, and died there in 1740. We have no particulars of the event, nor are there more than allusions to it in the journal of the date or in the letters of contemporaries. Lady Lavinia Pitt, however, mentions the disease as the smallpox, then so much dreaded.

He left no family—except his young ward, the mysterious daughter of "Mr. George"—of the Tretelly Inn. To her Antrobus had given his name, and she inherited half his estate. Shortly after her kind guardian's death she married an Exeter gentleman of high family. Her father, "Mr. George," died in the course of Mr. Antrobus's stay at Tretelly.

To some beaux and belles of the reigns of George II. and George III. this book, originating in the conversation of another George—George the Unknown—could well seem an interesting matter. All the more might it be so in view of its scarceness, from the first. There are no more copies of it, despite the fact that fashionable dilettanti in things occult have borne it in mind. Could anything be more characteristic of Horace Walpole than to find him in a letter, from serene Strawberry Hill, confessing—to no purpose—that he is "desirous of getting hold of that damned queer old woman's fortune-telling book, by Bob Antrobus." In the Diary of the sprightly Louisa Josepha Adelaide, Countess of Bute (afterward so unfortunate a wife and an even more unfortunate mother), she describes a droll scene at a Scotch castle one evening, in which the unexpected statements of "The Square of Sevens" as to the lives and characters of the company "put to the blush several persons of distinction" who rashly tempted its wisdom—especially including the aged Earl of Lothian. For what Lady Morgan thought of it, and the characteristic story of the peculiar terms on which she offered "to sell her copy to Archbishop Dacre," the reader is referred to the Bentijack Correspondence.

It is on its face a model method of fortune-telling with cards; easily the first for completeness and directness. Our author, in a letter to his cousin, Henry Antrobus, quotes the eminent Brough as styling it not only the most authoritative little book on its topic, certainly the most interesting one; hit the only volume on the subject "which is not a confusing and puerile farrago of nonsense—troublesome to look into and unsatisfactory to acquire." Certainly our ancient enthusiasts record can be learned and used systematically, exactly as is the case with such excellent and approved systems of chiromancy as Mr. Heron-Allen's and others. It may be thought fortunate for modern students of card-divination that the work has survived, so complete and clear. Its discreetness, too, is delightfully adroit, when it suggests that its tenses, past, present, and future, are not as definite as one might desire.

There is no copy of the hook in the British Museum, nor in the Paris Bibliothèque Nationale, nor in any public collection of America, England, or France that I can name. One worn but perfect MS copy is to be found in a private library in the United States. Another might yet be sought in far Australia, if still owned by descendants of Mr. Antrobus's young ward. Only by a special personal interest in the matter, and with a sense of risk to an heirloom, I am permitted to make the manuscript for this edition.

Undoubtedly, as "R.A.," Mr. Antrobus dressed the mystic "Significances" of the cards in the book's "Tavola" in English less blunt and uncultivated than they came to his ears from the lips of the dying "George—." But that he took no other liberties of the least consequence is pretty certain. He respected the "Supernaturall" here, as in his grave brochure on the Cock Lane Ghost, which spectre, alas! mightily took him in. And, by the way, the reader will please observe in his pages here following that though the method of "building" and so of forming the "Square," and of "reducing" it, seems at first glance bothersome and complicated, it is only a childishly easy performance in the way of making a square of seven rows of seven cards, and then of making the rows only three cards deep, at most! Crazy superstition and the aim at mummery have added the details of process that seem tedious. And, really, they are not ineffective in a drawing-room.

What we read of thus as carefully put together, conscientiously printed as a thing to be taken with seriousness, in its author's time, may in our social day serve a lighter end—and entertain the parlor, rather than awe the boudoir. With this intent, as well as in offering something of a literary curio, the present Editor assists it toward the glimpses of—not the moon, but the electric chandelier. And its Nineteenth-Century sponsor hopes that many curious and pleasant "fortunes" may be read by it; and that in its pages the ominous Spade, the mischief-working "Influencing-Card," the stern "Master-Card," the evil "Female or Male Enemy," and the "Vain and Amoratious Man" (who must be ever, indeed, a terrible combination to endure) may not be frequently encountered—in any case, that along with many other troubles and trials, such unpleasing meetings may not come outside the vagaries of a pack of cards.

E. IRENAEUS STEVENSON.

New York, 1896.

BRADAMANTE. But is this authentic? Is it an original? Is it a true, original thing, sir?

GRADASSO (*making a leg*). Madam, 'tis as authentic as very authenticity itself—
'tis truth's kernel, originality's core—provided you are but willing to believe it such.

BRADAMANTE. Sir, you quibble.

GRADASSO (*making a leg*). Madam, 'tis precisely in my vocation to quibble,—and delicately.

From *The Superglorious Life and Death of Prince Artius: A Tragedy*. Act LI., sc. li.

THE SQUARE OF SEVENS

Of the Preparing of the "Square of Sevens" from which is made the Parallelogram; with the due Shuffles, Deals, and Disposals thereto.

Take a Pack of Fifty-Two Cards, Shuffle the same well, Seven times. Then present the Pack to the Person whose Queries you seek to answer, who accordingly shall be called your *Querist*. Therewith must your *Querist* chuse from the Pack, without seeing the cards in it—three several Cards, which are to be called his *Wish-Cards*; the same being chosen with a Cut between each Choice. The *Querist* must not seek to see these same Wish-Cards; they are to be laid apart on the Table, or left to Repose in the *Querist's* care, till all that followeth of the *Square*, the *Parallelogram*, and the *Reading* be ended.

Of the Dealing of a *First Seven* Cards.

Again take in hand your Pack and Shuffle it yet smartly, there being Forty-nine Cards now left in it. Proceed next earnestly to Deal them forth on the table in the following Order and Manner, and without first seeing their Faces. And be solicitous of laying them down just as they shall come, Faces upward, in a Downward and Oblique Line; taking them from the Topmost of the Pack until you have laid forth Seven, Cards. And while you cruise and lay down the same, and indeed during all that here ensueth of Directions for your following, avoid foolish Conversation and sottish Pleasantries with those about you;

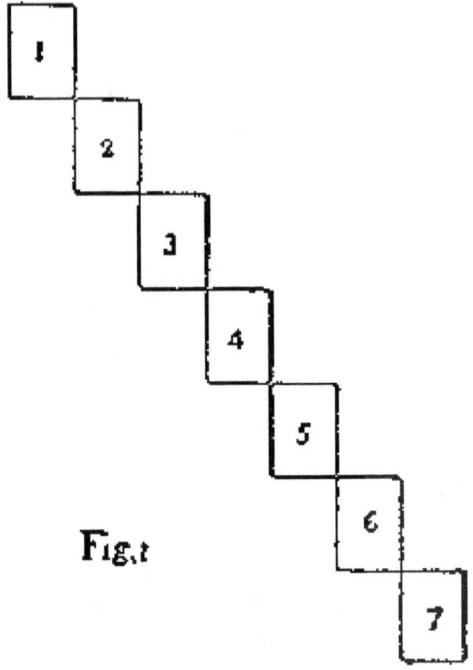

Fig. 1

having your mind serious to your task. (Fig. 1.)

Shuffle again; and therewith from the store of the Pack add to the above Seven Cards, a Dealing of *Six* more, to be taken from the Bottom of the

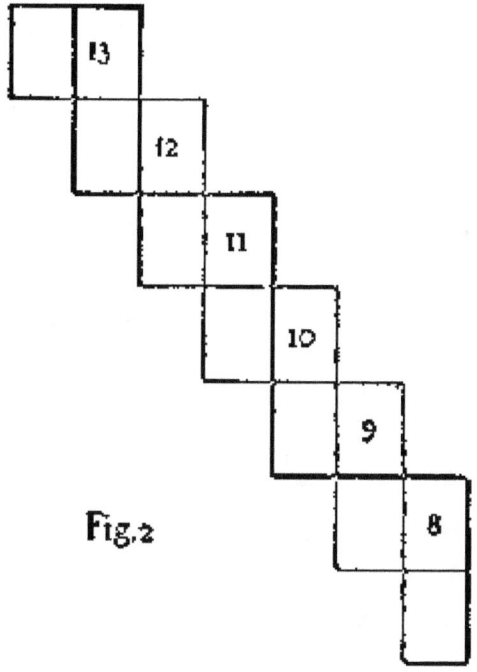

Fig. 2

Pack, chusing these also without knowledge of Suits or Values. They shall be laid in an ascending Border of *Six*, to the Right Hand of your first Series. (Fig. 2.)

Again Shuffle; and deal out

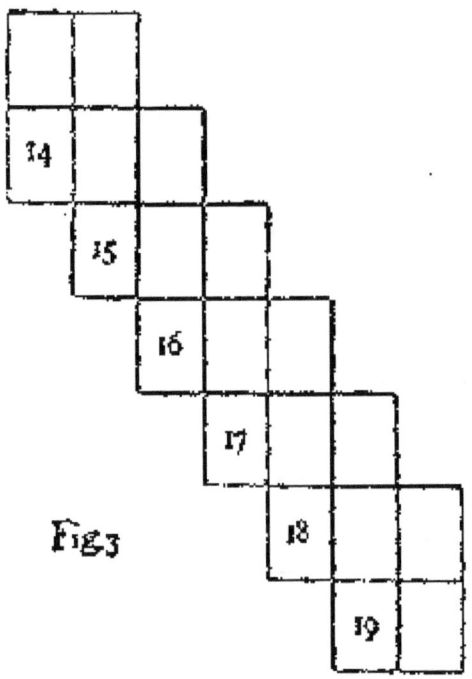

Fig 3.

from the Top of the Pack *Six* other Cards laying them in a downward Border, leftward, to the thirteen already placed. (Fig. 3.)

Nineteen cards now face you in the Series. Shuffle again, and deal from the Bottom of the

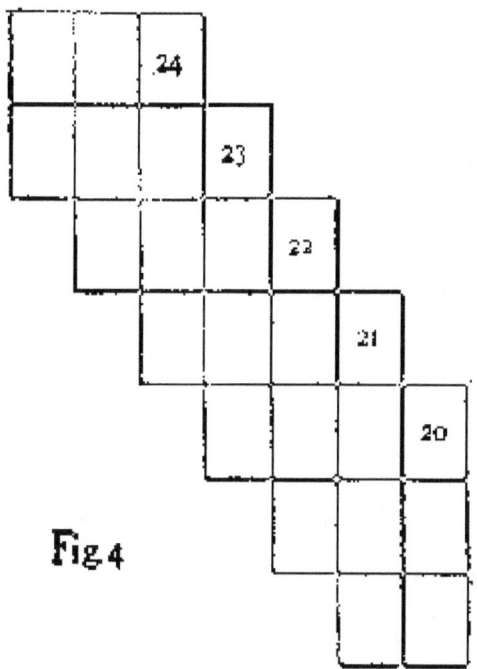

Fig 4

Pack, to the Right hand of the Figure a-making, *Five* cards, in an ascending Border. (Fig. 4.)

Shuffle again; and from the Top of the Pack deal down other *Five* Cards, laid in descent,

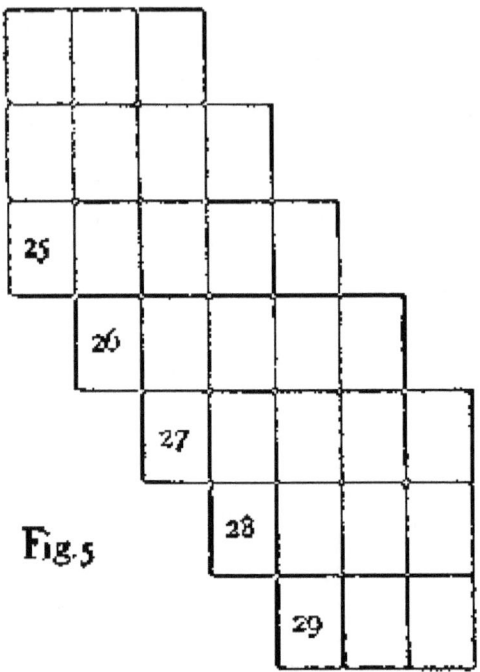

Fig. 5

as a Left-Hand Border. (Fig. 5.)

Your next Shuffling and Dealing will cause you to lay down, from the Bottom of the Pack, *Four* Cards, in a Right Ascension (Fig. 6), as you were laying

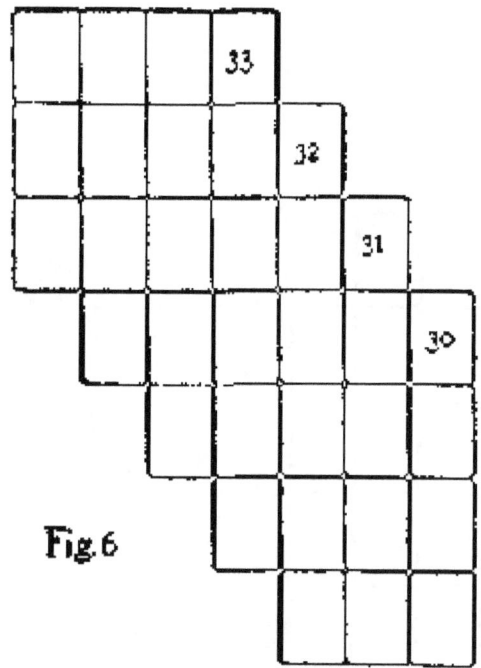

Fig. 6

a Tiled Floor or a Ceiling Pattern. There are now three-and-thirty Cards laid. Again Shuffle, and from the Top of the Pack lay—downward—a Leftward Border of *Four* Cards. (Fig. 7.)

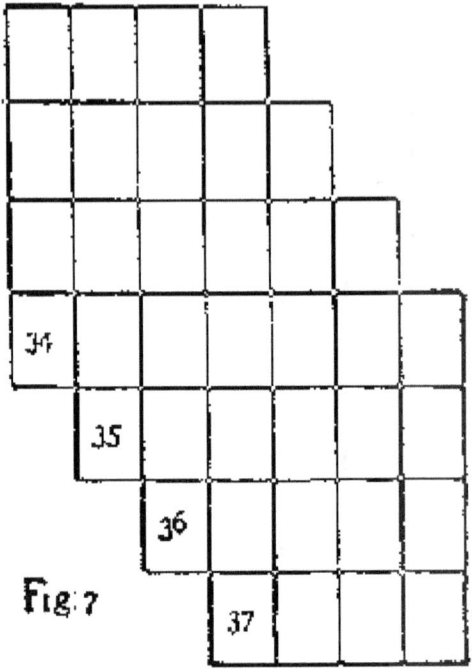

Of the sudden appearing of the Fair *Cross*.

Observe that each side of your Figure thereby has Four Cards, and its midmost Rows, horizontal and perpendicular contain, like the first Row laid, Seven cards apiece; and offer thereby a Fair *Cross*, the Goodliest Sign.

The first Figure needed, the *Square*, is now nearing Shape; its Rows, diagonal or horizontal or perpendicular, equalizing. Shuffle, and deal from the Bottom of the Pack, a Rightward Border of *Three* Cards, upward laid. (Fig. 8.) You have now four rows of sevens, in your figure: a *Cross*, withal still to be found in its middle—reckoned up and down and right and left.

You must now Shuffle—so well as your lessened cards will allow, yet with but one Shuffle—and deal, from your Pack's Top, a downward Left-Hand Border of *Three* Cards. (Fig.9.) Note that you now have in your Figure *three Rows of Sevens*; which you may well

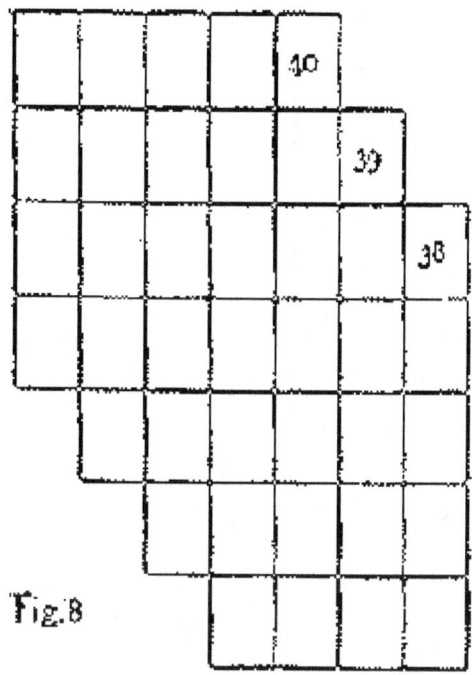

Fig. 8

wish were Guineas for your Purse.

Again Shuffle, as best you can, and from your Bottom Cards in the Pack deal *Two* Cards, laid at the right, upward, as in Fig. 10, nearly finishing

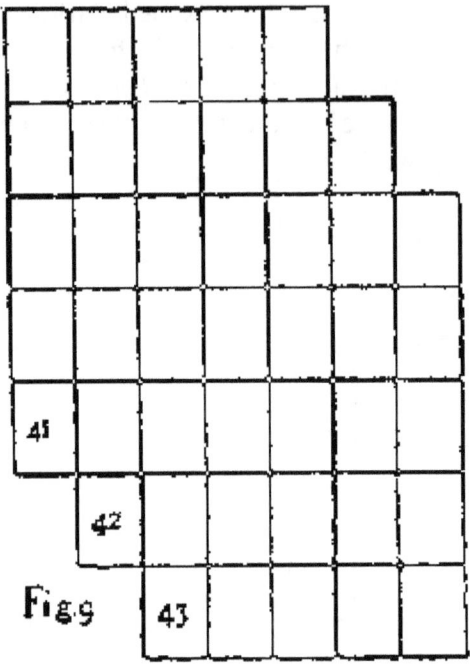

Fig. 9

your Square, now holding *four* rows of *Sevens*.

Again mix your cards; and deal from their Top, *Two* Cards, leftward laid in descent, as in Fig. 11. How much, O, Friend, in Human Life is nearly

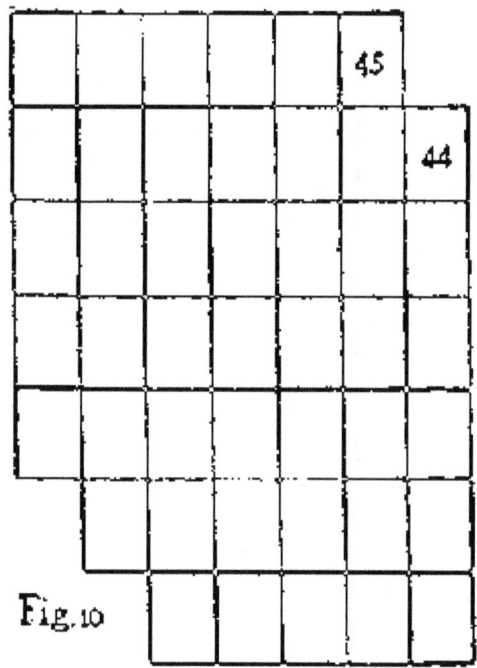

Fig. 10

perfect, yet not quite perfect!

Truly no Man shall reckon this Life a Perfect Matter with him! R.A.

Confuse now, as best and honestest you may (for you can hardly essay a shuffle), at least Once your two last Cards: and so complete your *Square*

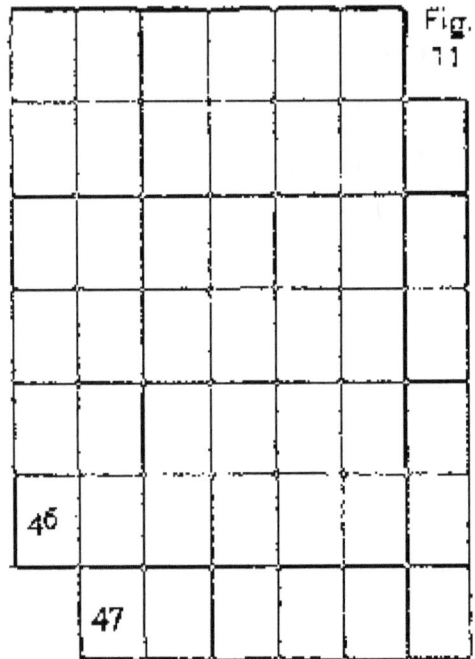

of Sevens, out of which will presently rise the Parallelogram; by laying its Forty-eighth and Forty-ninth cards in Opposition Corners, as in Fig. 12.

The *Square* is completed.

Consider this Figure attentively.

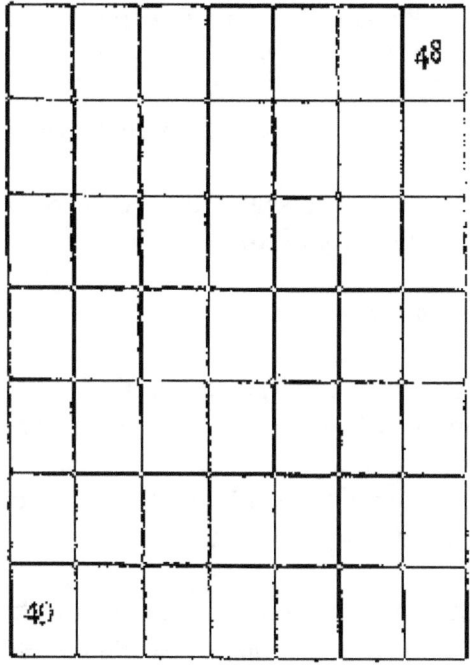

For now have you before you the very *Square of Sevens*: being a magicall Square Figure of Forty-nine Cards, whose Rows include ever *Seven* Cards, taken anyways. And that same mysticall Square now must be made ready for use in Reading your Querist's Fortune (or Experiences) by making it into a Parallelogram of smaller compass, through what is called its formal *Reduction*.

Of the Square Formally Reduced to The Parallelogram; and of The Master-Cards & The Sacrifice.

Having thus built your Square, chuse the extreme right-hand card in its Uppermost Row. Lay it on such card of the same Suit as lieth nearest it, in the same Row, if there be such; save on the last Card on the Left of the Row.

Of the *Master-Card* and *Master-Column*.

This Leftward Card in each Row is called the Row's *Master-Card*, and it cannot be covered by any other card, nor be moved. It must open and alone abide as it is; and the Seven Master-Cards, counted downward on the Square's left Edge, make what is called the *Master-Column*.

If you have no other card of the same Suit as your rightward card—or none save those of the *Master-Column*—let it lie. But if you can lay it on another, not a *Master-Card*, of any degree in the Suit—for observe the degree here matters not—so do. And then mark if, leftward and toward the *Master-Card* of the Row, lies another of the same Suit. If so, take up the two cards you laid together; and lay them on this third one. Look again and carefully; and, if another of the Suit be found, carry to it the former ones. So do until you have no more of its Suit toward its left, to join unto, and until all the cards of one Suit in the Row lie piled together; save the *Master-Card*, which even if it be of the Suit lieth alone, leftward, as it was first dealt out and down, in the last Square.

Of the *Reduction*.

Proceed then with the next Suit in the Row: and so with each Row, until you have thus sorted all the cards save the Master-Cards. This is the *Reduction*.

Close now and straiten-up together each of the Rows thus broken into Piles; pushing ever toward the *Master-Row*. Thus have you a new Figure, smaller than the last Square of Sevens, and somewhat irregular: there be in some Rows five cards, in others less; even so few, though rarely, as three or two. Note that a Pile of Cards is reckoned only as one card. Note, too, that with cards that have become hid beneath others you have no concern.

Of the *Sacrifice*.

Next, cast or lay aside in a parcel, all Cards in the Figure that are not contained in the three Columns, leftward (the *Master-Column*, and two other Columns). If, your *Reduction* done, any Row offer to sight but two Cards—the *Master-Card* and its neighbor—so must the Row abide. But this comes rarely. You will best not disturb the Cards hid in the Piles, for it is well to let Sleeping Dogs lie, and moreover needless Fingering and Quiddling of the Parallelogram is not commended. With the cards thus rejected have you no more to do. They are called the *Sacrifice*.

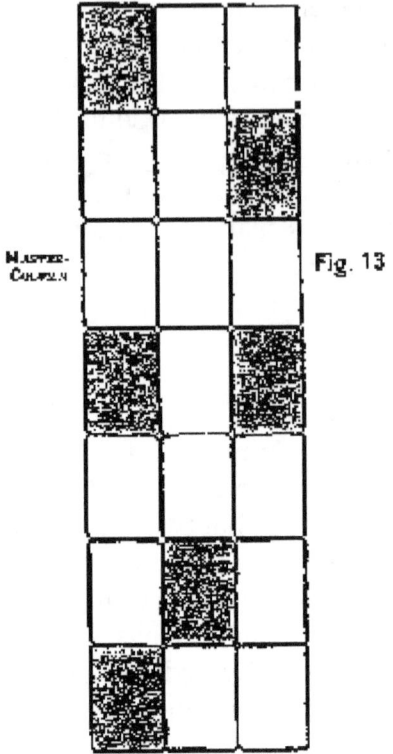

Fig. 13

The *Parallelogram* made.

Now have you a *Parallelogram* of One-and-Twenty Cards in sight (Fig. 13), reduced from the Square that formerly held Nine-and-Forty. With these One-and-Twenty now under your eyes I will be your Querist's affair.

You may indeed ask why so much Labour is made of building the Square only to reduce it, to despoil it, and to force it to hide or to part with so many of its Sevens—as by a sudden Slaughter or a Panic or a Plague. But it is held that by such prior Shufflings, Dealings, and Placings are much cherished the accidentall Declarings of Fates intelligence; and that by the other Processes, embracing The *Sacrifice*, there remain for *Reading* just the Cards decreed; free from disposition by light-fingered Craft, or from ticklish Arrangements by Skill.

A Thing of Great Mystery and Fair Harmony—as Jacobus of Utrecht calleth the *Soul*.

And the Square itself, the Parent of the Parallelogram, is of great Harmony as a Mystery. Indeed all other Methods of reading fortune in Cards are incomparable to it.

Of Summarizing in the Parallelogram its Aspect and of the Fortune or Experience of the Querist that it will Report.

With your Parallelogram thus built, observe it as an Whole; and remark if it hath an Agreeable or Unpleasing *Aspect*—one Auspicious or Unkind, according as it contains rather the red or the black Suits. For a Red Aspect is kindly. A Black Aspect contains many less favorable cards, especially if they be *Spades*.

Of *Hearts* as a Portent.

And, for another Matter, and a wider Notice as to the Suits of Cards:—it has long been assured by those best knowing Card Intelligences that the Suit of *Hearts* is the Suit of the Affections, Passions, Fancies and Feelings.

Of *Diamonds*.

And the Suit of *Diamonds* ever refers to Condition in Life, Society, Wealth, Position and the Fine Arts; and contains many Comfortable Cards.

Of *Clubs*.

In the *Clubs* lies the Judgment, the Intellect, the Will, and the Affairs of a Man's Brains, and what he doeth of his own Mastery and Genius.

Of the ominous *Spades* suit.

The *Spades* is ever the suit of doubtful or worse Prognosticks; of the Events that arbitrarily fall to Man's Lot, those things which hardly can any Prescience or Plans or Conditions of our own making amend. Thence is it that in especiall comes a serious, nay even a gloomy appearance to the *Parallelogram*. Your first Glance at it, therefore, gives you a Generall Character in it, to state first to the Querist before its details.

Of a particular *Uncertainty* in a Prognostick.

But particularly note that Matters to be read in its Cards may often refer not to the Future, or to the Present, but to the Past. Especially is this the truth with the Old or Elderly or with those Wed. Such must expect to be told of Experiences that lie behind them, rather than before them, of Good or Evil; for Fate oft allows sparingly of Incident to those of middle years, or later; and therewith she is often pleased to make her Oracle speak coldly to a Querist, of Ancient Circumstances.

The Shot seldom goeth twice into the same hole; and a Dead Trouble or Joy rarely Reviveth. And a Blessed Thing that 'tis so!

Hence, whether a Significancy in a Card speak of what is come or is yet to come, at best is none too certain; only it is true that the greater or harder Experiences of Mortall Lives seldom be duplicated. With the Young or Unwedded, the Significancies are of the Future, with far more determination.

Of the Reading of the Parallelogram, according to the Tavola; and of the Wish-Cards.

Note now your Card in the Right-Hand column, and also the Card next it, of course to the Leftward; which Leftward Card is spoken of as *Influencing* the other.

Of the Influences, In which the Philosophick will find a likeness to Human Circumstances.

The Significancy of it, for good or evil, is given in the *Tavola* that follows in this Book, by its proper Suit and Degree: and this you will tell to your Querist. Next note the card, which was just now an Influencing Card, but which, now in its turn, is to be considered according to the *Influence* cast *on* it by the Master-Card, beyond it, leftward. Declare this Significancy. Last, declare what may be the Significancy of the Master-Card, as such and alone. And so proceed, as to each card in the *Parallelogram*, ever naming *last* the Significance of the Master-Card, until your *Parallelogram* is all interpreted to the Querist. And note that the Master-Card even as an *Influence* is not more potent than another, (as far as is known), and that its Dignity and Potency arise only in its being uninfluenced; and, so speaking, from its Significance with a certain Individuality not belonging to its two Fellows. Nor are there; any Influences cast *Upward* or *Downward* by the Cards, out of the Row in which each lieth.

Having read the Parallelogram from beginning to end, slowly and honestly, lay forth those three *Wish Cards*, early chosen by your Querist, but not dealt in the Square.

Of the *Querist's Wish*.

If they contain more *Red* than *Black* Cards, this shall come: the Querist may wish a *Wish* for his own Profit or Pleasure, even in removal of an Evil that hath been read to him in the *Parallelogram*. If there be Black rather than Red cards in the Three, he must wish a like wish for *Another*. And in either case, if the cards deciding his Privilege be of high degree, such as *Court Cards*, Aces or above the Eight, his Wish is likely to be granted, or at least it is not in vain in some sort. But if the Cards be low in Values he has desired to Fruitlessness.

Let it be minded that by the Phrase an *high* or a *low* Card in a Suit is ever meant, respectively, the cards above or below the *Eight*; the Aces being reckoned as the highest in a Suit. And indeed Cards must ever be read with a Considering of their Degree, and of the Six in Court Cards. Where there be no speciall Significancies given to the Degrees, the Judgement must I shift as best it can.

It is well not to oblige, of any one evening, or on a set and single occasion, more than a Querist or so—maybe, oblige at most three Querists—by making Squares of Sevens and Reducing the same and Reading what may lie therein. Too much of any good thing makes it over-common, blunts the Appetite and dulls the Apprehensiveness of the Reader. With fatigue, too, may come Carelessness and, on good occasion, even Lying: and, besides, let us respect the Supernaturall.

THE *TAVOLA* OF SIGNIFICANICIES AND INFLUENCES, PROPER TO THE TRUE *READING* OF THE PARALLELOGRAM; ADJUSTED IN A SYSTEM OF ALTERNATIVES

NOTE

In transcribing this "Tavola," the Editor has somewhat modernized the spelling and capitalizing, for the convenience of the reader. With reluctance, but of necessity, he has also amended—or emended—the phrasing, where it is in the original hardly consonant with modern taste.

OF HEARTS

The Ace.

As a *Master-Card*, a special Emotional Experience. Influenced by a *King* of like Suit, there is figured an Intimate Friend, or one in whom the Querist is much bound. By a *Queen* of like Suit—an Emotion for a Woman of beauty and charm. By a Knave of like Suit, an Attachment to a Man younger than the Querist. Influenced by any *high heart* other than those above, an Amorous or Affectionate Temper of mind or body. By a *low heart*, an impressionable, kindly Nature. These are Five Special Interpretings. The more general are: influenced by a Diamond, Good Fortune in something, measured by the degree of the Influencing Card. By a Club, a Talent or Gift to be made much of. By a Spade, an Error, or Disappointment, in the degree of the influencing card.

The King.

As Master-Card, is figured that the Querist deals or has had much to do with a Man of fair skin and light type, of good temperament. Influenced by an Ace of like suit, one notably unselfish. By Knave, a Lover, Husband, Friend. By a Queen, a Love-match. By a Diamond, a Man of Wealth or artistic nature. By a high club, a Man of Energy withal; by a low club, one of Prudence. By a Spade, a man of some defect of Temperament, or of a Chronic Malady or Blemish, ominous to him and others.

The Queen.

As Master-Card, is referred to specially, an amiable, affectionate Woman, rather one sentimental than of intellect. Influenced by like suit, if an Ace, she is admired of Many; if a King, she is wedded, betrothed, or beloved by one in especial. By a Knave of like suit, she is beloved by a Male Relative in especial, not of her own near family. By other cards of like suit, degrees of regard. By a Diamond, a Woman gifted, and esteemed much in Modish Life. By a Club, though not learned she appreciates knowledge in others. By a spade, she is not of firm health; or not of wholly firm Virtue.

The Knave.

As Master-Card—the Querist's closest Friend; yet likely held such because of feeling rather than judgment. Influenced by an Ace of like suit, there is no Inequality in the affection. By a King of like suit, Resemblance to the Querist in physique or mind or disposition. By a Queen of like suit, one with distinctively feminine traits. By another card of like suit, a popular man with his fellows. By a diamond, of wealth or social Position; but if by a Nine of

Diamonds, not enduring in such Happy Fortune. By a Club, a Friend of judgment and good at advice. By a Spade, a Friend of not too sound health: or apt of offence.

Ten.

As Master-Card, a general reference to Matrimony, as being ever the card-matrimonial. Influenced by like suit, a High-Marriage and that auspicious: by a low heart, a Marriage not one's first or first-wished. By a Diamond, a Marriage with money in it. By a Club, a Marriage of reason or of circumstances. By a Spade, an Interrupted or more or less Disastrous Match.

Nine.

As Master-Card: a Card of Good Augury for what we wish for Another. Influenced by its like suit, an unexpected Meeting, with a person much affected or desired. By a Diamond, a Pleasure in store. By a Club, a Wish partly fulfilled, rather than wholly. By a Spade, a Wish fulfilled but followed by some detrimental Event.

Eight.

As Master-Card, a Love-Interest. As influenced by like suit, an Interest of much Romance. By a Diamond, a Lost Article recovered. By a Club, the Victory in a difference or argument as to some plan or act. By a Spade, a Caprice to warm the heart; or a new Article of dress or household stuff.

Seven.

As Master-Card: the Card of Trust and Confidence approved of. Influenced by like suit, honest Love, or

Family regard. By a Diamond, wise Trust in a commercial or social step. By a Club, in a Secret. By a Confidence misplaced in a person or event.

Six.

As Master-Card: A strong Inclination, a Desire, or Action is well rewarded. Influenced by like suit, it concerns another even more than ourselves, or as much. By a Diamond, a step of social or artistic or pecuniary vantage; save if the diamond be the nine, which leaves the result in Doubt of full success. By a Club, a Matter of Judgment and practical bearing, seen and discussed of others; or a Remark, or a Letter of more consequence than would appear. By a Spade, an Inclination or desire, not wholly honorable: or of brief realization.

Five.

As Master-Card, an amusing and diverting Affair heard of, or entered into. Influenced by its like suit, a Feeling not hitherto returned is met at last. By a Diamond, a Success in some-thing particularly wished. By a Club, a keen and shrewd Chance at a remark to be well caught. By a Spade, an Ache, Pain, or Breaking.

Four.

As Master-Card, is figured the existence of an obstinate Sentiment toward one, or an Opinion not of our own building up. Influenced by the like suit, it is troublesome, causing thought, new to one, or burdensome. By a Diamond, it is known to others, or guessed. By a Club, it is apt to lead to acts officious or of manoeuvre. By a Spade, it is a Sentiment based on error and lack of full insight; or it will be abruptly weakened.

Three.

An Act of Charity and Generosity, by or toward the Querist, if read as Master-Card. Influenced by like suit, Action in a matter of very confidential sort. By a Diamond, it is in part a Matter of Money or Office or from a Superiour—and may be associated with an investment, a society, an entertainment. By a Club, it figures a Visit, or Visitor. By a Spade, a Change of Opinion in some near matter is enjoined, or the Loss of a good will; or a Surprise not welcome wholly.

Two.

As Master-Card, favorable News, or a Letter acceptable. If influenced by its like suit, the Person from whom it comes, or also referred to in it, is much valued, or a near Relative. By a Diamond, a Present, a Visit, a Meeting of service, a Letter, respectively. By a Club, a "yes" in a matter open. By a Spade, it concerns Another more than the Querist; or else will not be altogether correct in statement.

OF DIAMONDS

The Ace.

As Master-Card, a tangible and material Success in some Matter of Society, Money, Art, or Office. Influenced by a King of like suit, a Loss recovered. By any other card of like suit, Information and certainty of an Affair of purchase, bargain or sale, much to advantage. By a Heart, a wise Marriage, the settlement of a Difference, an open matter closed to satisfaction. By a Club, a prudent Choice. By a Spade, a Cost or expense, perhaps a loss, before a satisfactory and favorable Event, or in course of it.

The King.

As Master-Card, is figured a brilliant, honorable and successful Man, of standing and perhaps of marked taste in art, belles-lettres and the like; and gifted in them. Influenced by its like suit, a Man with much original in him, shrewd in money or gift. By a Heart, a Male Character of kindly and humane traits; or one sensitive and easily moved in his mood. By a Club, a Man in professional life, and of good mental balance. By a Spade such a life is threatened or broken, or not free from Self-seeking at others' expense.

The Queen.

As Master-Card is indicated the existence of a brilliant, gifted Woman; fond of social life and modish things, of dress or expensive and rare matters; perhaps of Talent in art or literature. Influenced by like suit, one of brilliancy rather than feeling or self-sacrifice. By a Heart, if high, of affection more than is thought; if low, beautiful. By a Club, a Woman executive; of some audacity; restless or self-depending: admiring intellect of solid kind tho' maybe lacking it. By a Spade, a Woman not devoted to benefiting others; and threatened by misfortune; or with a hidden Grievance.

The Knave.

As Master-Card, is figured as within the Querist's life, a Relative, likely so made by birth or marriage; and ever disposed to use the tie for personal advantage. Influenced by like suit, the Relative is not remote, and marriage or love is so utilized by him, now; especially by weakness of judgment, or by over-affection on another's part. By a Heart, a shrewd Business Success. By a Club, a sudden

Discovery as to a person. By a Spade, a Deferment of the Querist's prosperity in a matter.

Ten.

Also, as Master-Card; a brilliant, entertaining, but too trifling and irresponsible Man: or a vain and amoratious man if a knave of beads influence it, often is figured.

Nine.

As Master-Card, a valuable Possession. Influenced by like suit, is concerned one intrinsically of value, as jewels, money or plate, a house or estate. By a Heart, a Secret: a Marriage. By a Club, the aforesaid or another Possession will be (or has been) won by special exertions of the Querist's abilities, or so to be kept. By a Spade, it is endangered.

Eight.

This is the Unlucky Red Card if figuring as Master-Card; meaning a personal Event of importance going awry; a Subtraction that must be admitted to others. But if influenced by like suit, it is a favorable card and indicates a pleasing Journey, or Meeting. By a Heart, an Enemy or evil opinion altered in your favor. By a Club a Proposal of tempting kind. By a Spade, a Plan that in success is doubtful and partial, or troublesome to another.

Seven.

A card of good omen if a Master-Card, in the Practical Affairs of life, business, society, or art, or one of them. Influenced by a like suit, in a Commercial thing; a Meeting wished; an influence desired. By a Heart, a wealthy and superior, or happy Marriage. By a Club, a Communication of importance or good. By a Spade, an Indiscretion that were better not committed by your fault; or a Negligence.

Six.

As Master-Card, the card of special Report, Conversation about one, or of Action by another; in a degree affecting one's outward affairs. By a Heart, from a near Friend. By a Club, where you esteem or respect. By a high spade, with error or even untruth in it, mayhap not intended, but a pity. By a low spade, it is somewhat written.

Seven.

As Master-Card, a commercial or social Step, a Purchase of importance; by the Querist. Influenced by like suit, attractive and unexpected. By a Heart, in regard to making a new Acquaintance, or bringing a Change of feeling toward some one. By a Club, a Matter of Necessity; or an affair dealing with a lawyer, doctor, clergyman, or servant: or a Step of wisdom as well as attraction. By a Spade, if high, a Loan of money: if low, a small Borrowing.

Five.

A good omen; as a Master-Card, meaning a Gift to the Querist. Influenced by like suit, is figured a personal Ornament or convenience. By a Heart, a Gift is to be made. By Club, it comes with formality and after debate, and considering for some time, or for special circumstances. By a Spade, a Disappointment to another dear to you, is figured.

Four.

As Master-Card, an Honor or Favour or Compliment or bit of Luck. Influenced by like suit, in society, or art. By a Heart, long desired; and perhaps more pleasing than wise or useful. By a Club, due to one's own judgment and persistency. By a high spade, entailing trouble or cost. By a low spade, at the cost of another's misfortune; or not wholly our desert rather than another's; or brief.

Three.

As Master-Card, a sudden Surprise in an event. Influenced by like suit, agreeable, and social or pecuniary or in the arts. By a Heart, Surprise, agreeable, yet not to one's interest or particular profit. By a Club, a social Responsibility. By a Spade, a Death or a Misfortune to another likely enters into it.

Two.

A gift or fortunate Purchase, if a Master-Card. Influenced by the like suit, an Engagement or Burden happily broken or dismissed; a Good Riddance, a Disgrace or Plague ended. By a Heart, an Offer—in love, friendship, trade, travel, profession, or pleasure. By a Club, a Letter or Interview of consequence. By a Spade, a Service that one is glad of, or a Gift; but bringing obligation with it, sooner or later.

OF CLUBS

The Ace.

As Master-Card, is figured an Event of material weight, involving use of judgment, will, shrewdness, or decision. Influenced by the like suit, high or low, its effect is the more for our own making. By a Heart, is seen a Matter in which our Sentiments are specially enlisted, perhaps in contest with judgment or tastes or duty. By a Diamond, the affair is in society, artistic life, money, or responsibility to others as well. By a Spade, a Mischance or Disappointment is part of it; often faithfully hid, or to be hid.

The King.

As Master-Card, our relationship to a strong mental or moral Influence of the male sex, respected and deferred to; or sure so to be. Influenced by its like suit, it is a cultivated and professional one, or involuntary. By a high heart, it arises in a near relative or one for whom a special affection is felt. By a low heart, it is either secret or remote; or it may be that it is religious, in part. By a Diamond, our outward life must have concern in it. By a Spade, the influence is of doubtful or worse healthfulness or profit to us.

The Queen.

As Master-Card, a marked female Influence on the Querist, in the way of respect, judgment, advice, or authority: not necessarily as to a relative. Influenced by the like suit, a person of coldish and grave disposition. By a high heart, of strong impulses and disinterested; by a low heart, troublesome, often importunate and officious. By a Diamond, not married; and of wealth or social esteem; talented. By a Spade, not altogether open or disinterested;

divorced or disappointed; according to the nature of the Card.

The Knave.

As a Master-Card, Relationship with a well-meaning, but over-rash and hasty or sanguine Man; not necessarily but likely quite youthful, and selfish in inclination, or too easily influenced by others of greater art: an Associate, partner, friend, or Employee in some matter of worth. Not to be relied on as one would gladly do. Influenced by his like suit, Circumstances assist him or make of less or more account his weakness or strength. By a Heart, he is inclined to be led by tastes and passions and by skilled flattery, or to overtrust. By a Diamond, he is in love with externals, fond of dress, or notice, or pleasure; ambitious. By a Spade, he meets with Losses to himself and the Querist, or he makes some particular Error or False Step.

Ten.

As Master-Card, Success in a matter long pursued. Influenced by its suit, one of troublesome Conflict of conduct or advices. By a Heart, in an affair of love; or calling for courage; or for another, as well as oneself. By a Diamond, an Opinion or Prejudice overcome in others, through our persistency, or argument. By a Spade, an Inheritance; or a Matter needing much watchfulness and care, when known.

Nine.

As Master-Card, the need of much Decision in our own judgments in an affair of importance; a need of disregarding counsels of Others. Influenced by the like suit, several persons or circumstances Oppose, perhaps slyly. By a Heart, there is a wounding of tenderer feelings or relationship in it. By a Diamond, the affair is of Estate, Position, Money, Comfort, or Purchase. By a Spade, beware lest so is assumed no greater Responsibility than can be easily carried; or acknowledged.

Eight.

An absent Friend reflects on you in a particular matter. This as Master-Card. Influenced by like suit, a Conviction or responsibility of much weight laid on one. By a Diamond, a Choice of a wife, or precious article. By a Heart, Cause of Concern for a friend. By a Spade, you shall give Counsel not followed, and spend Thought thrown aside.

Seven.

As Master-Card, a troublesome Situation dissolved. Influenced by like suit, a Secret imparted of interest and length. By a Heart, undo something very newly done. By a Diamond, beware of an Indiscretion or Error. By a Spade, a Neglect or piece of forgetfulness will be of cost to mend or replace: perhaps, if a high spade, not to be mended at all.

Six.

As Master-Card, cancel at once an Agreement, a Purpose, or wholly change a Decision. Influenced by the like suit, a card of fortunate aspect. By a Heart, the call to assist Another, near to one. By a Diamond, a Hazard, successful. By a Spade, a sudden Opposition.

Five.

As Master-Card, a Guest, a Visit, a Letter, each needing exercising of prudence or self-restraint; but acceptable.

Influenced by the like suit, a Proposal urged. By a Heart, a Wound or Bruise. By a Diamond, a strong Temptation, or a Journey. By a Spade, an Argument or Dispute on a matter.

Four.

An important Request of the Querist, if read as Master-Card. If influenced by like suit, one not overmuch to your wish. By a Heart, you Sacrifice somewhat to grant it. By a Diamond, it involves anon a Change. By a Spade, the cost will not be valued for its worth.

Three.

As Master-Card, a sad or serious Duty or Care. As influenced by its like suit, a Choice of two things; both desired much, but one to be dismissed. By a Diamond, Luck, or a forthcoming Pleasure. By a handsome Man or Woman to be met and attracted toward one. By a Spade, a Matter to make one angry, or heart-sick.

Two.

A card of doubtful omen when a Master, figuring a grave Confidence, of interest to learn, but burdensome rather than easily to be passed by. By its like suit, some News. By a Heart, a Sentiment not wise though keen. By a Diamond, an awkward Meeting. By a Spade, a Piece of News acted on, and then found untruly reported: or Advice seemingly good, but not so.

OF SPADES

(*The Suit of Evil Omen and of Unwelcome Influences.*)

The Ace.

As Master-Card, the Ace figures a special Misfortune, Unhappiness, or Hurt to one's life, by no means avoidable, and perhaps not discernible at once. Influenced by the King of its like suit, sudden: by the Queen, long continuing ere complete; by the Knave, Fortune through Persons; by the ten, through concurrence of sundry events. By any other spade, sudden. By a Heart, Ill-Fortune in the Affection. By the Diamond, in the eye of Others, in society, money or art. By a Club, to our fear.

The King.

As Master-Card is figured a particular Man, our enemy, resolute and powerful. Influenced by its suit, it signifies News of a Death: or of misfortune to others. By a Heart, it involves abuse of Trust or Affection. By a Diamond, is figured a Man of social station and wealth or talents. By a Club, a Man cautious and reserved, and hence perhaps unsuspected for his real Malevolency.

The Queen.

As Master-Card, a Female Enemy, evil wishing or evil-working. Influenced by the like suit, known or soon to be shown as such to you, and the work. By a Diamond, comely and clever or gifted. By a Club, intellectual and audacious. By a Heart, her enmity arises in jealousy or vanity or in revengefulness or natural malice.

The Knave.

As Master-Card, a Man having no love for you and inclined to wrong and hurt you; but happily limited in Opportunity. Influenced by like suit, often seen of you. By a Heart, abusing your Trust, smaller or greater. By a Diamond, adroit rather than bold. By a Club, cruel and slanderous.

Ten.

As a Master-Card, an Event or Project to your disadvantage and regret. Influenced by the like suit, a Disgrace. By a Heart, a Quarrel. By a Diamond, a Cheat. By a Club, a Hindrance.

Nine.

As Master-Card, a Lie, or an unwelcome Meeting or Visit. Influenced by like suit, if a high card a Lie; if a low, a piece of undesired News or Letter. By a Heart, a sudden Alarm or Anxiety. By a Club, a Broken Promise: or a Secret told.

Eight.

As Master-Card, an Illness. By a high influencing card of the suit, a long Illness: by a lower, a shorter one. By a Heart, an illness to Another dear to one. By a Diamond, a Misfortune in an affair. By a Club, an Accident.

Seven.

As Master-Card, a tempting Proposal that must be declined. By the like suit, a Sharp Quarrel. By a Diamond, a Risk not welcome. By a Club, a Disappointment in a person or thing or event. By a Club, one arrives just too late for a certain Pleasure or Good.

Six.

As Master-Card, a Disappointment. Influenced by its own suit, a Journey not of pleasure, or else unpleasant. By a Diamond, a fall. By a Heart, a mistake of inconvenience. By a Club, must be read an unfavorable Sign.

Five.

As Master-Card, an Expense. Influenced by its own suit, a Neglect. By a Heart, a Worriment or Grief. By a Diamond, a doubtful Success. By a Club, a Death heard of.

Four.

An unfavorable Master-Card, affecting some near Concern to the Querist; belike it shall end less well than was hoped. Influenced by like suit, a Separation not welcome. By a Heart, a capricious Change of inclination. By a Diamond, a Perplexity. By a Club, a Loss.

Three.

As Master-Card, a suddenly changed Plan, a Discomfiture. Influenced by its like suit, a loss. By a Heart, a sudden Failure, a Doubt or Fear. By a Diamond, a Breach or Quarrel. By a Club, a sheer Folly, not to be warned away by a friend.

Two.

As Master-Card, you must say "NO," when you would say "YES." Influenced by its like suit, a Displeasure. By a Heart, an Evil Habit to burden. By a Club, a Strong Effort of no use. By a Diamond, a Folly, or a Mare's Nest.

In this evil suit of Spades there be many other special Significancies; but they are not pertinent to this method.

THE END

www.ingramcontent.com/pod-product-compliance
Lightning Source LLC
Chambersburg PA
CBHW061520180526
45171CB00001B/257